HUMBLE PI:

THE MATHEMATICS EDUCATION OF GIRLS

Muriel Eddowes

LONGMAN FOR SCHOOLS COUNCIL

1983

Published by Longman for Schools Council
Longman Resources Unit
33-35 Tanner Row
York YO1 1JP

CONTENTS

FOREWORD

This pamphlet is the first direct result of a joint venture between the Schools Council and the Equal Opportunities Commission. The Schools Council is charged with developing the school curriculum, the EOC with the promotion of sex equality; in 1979 they decided to pool their expertise in these areas and to collaborate in a series of pamphlets on sex differentiation in a range of curricular areas.

The Schools Council has affirmed its commitment to equal educational opportunities for girls and boys. It has set up a Sex Differentiation in Schools Working Party, and has funded a range of activities aimed at reducing sex bias in education. The issue of sex differentiation is increasingly recognized as important. Research continues to demonstrate the sexual inequality of our educational provision and many educationists are actively involved in developing ways of reducing the disadvantage experienced by girls and by boys during their schooling. The subject of this first pamphlet - mathematics - is particularly timely. The value of mathematics as a central component of general education and as a passport to further and higher education and employment is not contested. As a country, we can compete in industrial and technological terms only if we have a workforce well educated in mathematics. The Cockcroft Report* explored mathematics education in schools and commented on the relatively poor mathematical achievement of girls as a group.

This pamphlet considers the problem, outlines some of the reasons and offers practical suggestions for remedying the situation. The strategies outlined are sensible and realistic, draw upon the experience of practising teachers of mathematics and do not involve massive injections of resources. If girls succeed in developing confidence and competence in mathematics, they will be better equipped to succeed in a world of equal and changing opportunities.

LESLEY KANT
Schools Council

---

* Department of Education and Science, *Mathematics Counts*, report of the Committee of Inquiry into the Teaching of Mathematics in Schools (Chairman: Dr W.H. Cockcroft), (HMSO, 1982).

ACKNOWLEDGEMENTS

I would like to thank Gaby Weiner for her substantial contribution to
the text of this pamphlet, and Lesley Kant for her critical comments
and advice.

MURIEL EDDOWES

1. WHY LEARN MATHEMATICS?

Skill in mathematics is regarded as a basic qualification for transfer
between stages of schooling, and between school and work. At the same
time mathematics is seen as 'difficult' and 'boring', and success as
requiring a person with special talents - a keen analytical mind and
objectivity. Tradition does not normally confer these characteristics
on women.

However, there is also an alarming proportion of men who profess,
often cheerfully, their dislike of mathematics and lack of competence
in it - an incompetence which seems to be socially acceptable, unlike
deficiencies in reading or writing.

There is a dichotomy here. Mathematics is regarded as part of the
passport to progress in our working lives, but poor performance is
explained away by the subject's difficulty and its lack of relevance
to the real world. The problem is exacerbated for women in that they
have been conditioned to have lower educational expectations in general
and to assume they lack the attributes required for mathematical success.

If mathematics is difficult and uninteresting to all but a few, why
do we insist that everyone tries to learn it, and make success in it so
important a qualification in education and employment? The ability to
overcome obstacles seems rather a blunt instrument for selection.

The number of people who will learn mathematics for its own sake is
inevitably small. Not many of us learn a foreign language for its
aesthetic quality but rather to communicate in the language. Similarly,
most pupils should learn mathematics to be able to use it.

The utility of arithmetic, for example, is obvious. Most of us need
to be able to check bills, estimate mortgage and hire purchase repayments,
measure accurately, etc. With a pocket calculator you can obtain the
correct answer to a calculation, but you still need to know which
arithmetic processes to adopt and in which order.[1]

More erudite aspects of mathematics may seem to require more justifi-
cation to many pupils, but many mathematical techniques are fundamental
tools for other subjects - the sciences, design, technology, engineering,
geography, accountancy, economics, business studies. Pupils who progress
to further and higher education are frequently dismayed at the extent
and depth of the quantitative techniques required for their chosen course.[2]

Mathematics (as with other school subjects) is a valuable tool to help the pupil develop personal and intellectual skills. An analysis of these skills and qualities in relation to mathematics is given in a pamphlet published by the Mathematical Association.[3] The 47 skills given may be grouped into five broad categories. The study of mathematics should enable the pupil to develop:

(1) analytic skills - problem formulation and the selection of suitable solution techniques

(2) logical thought - inductive and deductive reasoning processes

(3) communication skills - precision and the use of symbolism

(4) skill in the application of a rule - both to solve problems at school and later at work, and the ordinary problems of living

(5) creative and imaginative thinking.

Teaching mathematics at school can be justified on many grounds - social, vocational, philosophical and cultural, and each view suggests a variety of aims and objectives for the learning process.[4] In order to achieve these aims and objectives, we need to think very carefully about the content of mathematics courses and to distinguish between requirements for arithmetic and mathematics. There is no point in teaching children mathematics unless it is done in such a way that they have the best opportunity of realizing their full potential in that area.[5]

The reasons for studying mathematics given here are equally cogent for boys and girls, and it is highly desirable for both sexes to be encouraged to study mathematics as one means of personal development.

2. MATHEMATICS EDUCATION: RELATIVE INVOLVEMENT OF BOYS AND GIRLS

Some important points about sex differentiation in mathematics involvement were revealed in an HM Inspectorate survey in 1973-74.[6] Table 1 gives a summary.

Table 1   Composition of mathematics courses 1973 (%)

	No maths	Some maths	CSE	O level	6th form non-exam	O+*	One A level	Two A levels	Post A level
Boys	14	49	48	54	56	51	73	82	89
Girls	86	51	52	46	44	49	27	18	11

Source: Based on data in Tables 8, 9 and 10 in Department of Education and Science, *Survey of Mathematics in Schools 11-18 (1973-74)*, (DES, 1975).

More recently, Bibby's study of the pattern of choice adopted by Scottish pupils[7] showed that at each decision point through secondary school more girls than boys drop out of mathematics, until by the post-school stage only 4 per cent of girls remain compared with 11 per cent of boys. Bibby found the largest percentage drop-out from the SCE examination course was thirteen-plus, the withdrawal rate being 19 per cent greater for girls than boys.

Several social and economic trends have led to a large increase in the numbers of pupils staying on into the sixth form.[8] This increase is reflected in the overall numbers of GCE A-level candidates (see Table 2); the total entry has risen by almost $4\frac{1}{2}$ times for the boys and by $7\frac{1}{2}$ times for the girls. Although the percentage of female entries in mathematics has risen slightly, the percentage of male entries has remained virtually constant, leading to an overall small decrease in the total percentage of A-level candidates.   There are, however, still three times as many boys as girls entering the subject at this level.

---

* Includes O-level retakes, O-level Additional Mathematics and O/A-level Mathematics.

Table 2   Candidates entered for pure and applied mathematics A-level examinations in summer 1951
and 1979 compared with entries for all subjects

| | Summer 1951 | | | Summer 1979 | | | % of total entries | | | |
	All subjects	Maths	English	All subjects	Maths	English	Maths 1951	1979	English 1951	1979
Male entries (in OOOs)	71.50	9.58	6.15	312.21	41.13	19.17	13.4	13.2	8.6	6.1
Female entries (in OOOs)	32.30	1.31	6.78	242.69	13.46	41.10	4.1	5.5	21.0	16.9
Total (in OOOs)	103.80	10.89	12.93	554.90	54.59	60.27	10.5	9.8	12.5	10.9

Source: Department of Education and Science, *Statistics of Education 1951 and 1979*, Vol. 1 (HMSO, 1953 and 1981).

A consideration of the corresponding figures for English allows the position for mathematics to be put into perspective. Competence in English is also important for all of us. At A level the girls predominate. Between 1951 and 1979, the rate of increase in entry is similar to that for mathematics, but the proportion of entries has dropped for both sexes. However, the representation of the girls in this area is still much greater than in mathematics; the reverse being true for the boys.

The increase in size of sixth forms has undoubtedly produced an increase in the number of girls involved in advanced study, including mathematics. However, the proportion of all pupils entering for mathematics at A level is not commensurate with the upward trends in other areas. There may be many reasons for this. Mathematics may be regarded as too difficult a subject by the new sixth formers; the increased provision of other courses may be distracting potential mathematics students; it may not seem relevant to future needs; there is also a wide range of non-examination courses now offered in many schools - 'mathematics for science', 'statistics for geographers', etc. - and these subject-specific support courses lessen the need for the more rigorous and less directly applied A level course.

These two trends - that the proportion of girls involved decreases rapidly as the level of the mathematics course increases; and the relative decrease in the proportion of all pupils studying mathematics for public examination after sixteen - have far-reaching implications. There is an inevitable shortfall in the number of professional mathematicians, and fewer pupils progress into mathematics-related areas - for example, science and engineering.

Part of the problem lies in early curricular choices. Unless there is thorough counselling and careers advice when options are chosen at the age of thirteen or fourteen, many pupils, both boys and girls, close doors through which they may later wish to pass. If a girl, for whatever reason, chooses history rather than geography, a language rather than physical science, needlework rather than woodwork, etc., her apparent need for a high level mathematics course is reduced. (Many options, such as woodwork, may not be offered to girls in the first place, either because of timetable difficulties or school policy.) The difficulties arise at sixteen when she finds herself debarred from many careers or further education courses because of inadequate qualifications in mathematics or science. In this way many girls are channelled into traditional jobs or courses.

How does the withdrawal of girls from mathematics courses compare with their withdrawal from the education system as a whole? For girls who persevere with mathematics, how does their success rate compare with that of boys?

The problem in mathematics is partly a reflection of the general withdrawal by girls from full-time education, but the drop-out rate for mathematics is much higher than the overall rate.[9] This is consistent with the differential career expectations held by parents, teachers, employers and the girls themselves. Girls have much lower career expectations than boys of comparable ability,[10] and tend to go into traditional areas of work which used to have no major mathematics requirement. However, more and more employers in these

jobs are now asking for more mathematics qualifications.[11] The
changing nature and size of the labour force, together with the
dwindling job market and rapid invasion of more advanced technology
into all fields of work, demand a much higher level of numeracy and
competence in basic mathematics than ever before. This is true for
both boys and girls. Although the problem is more acute for girls,
many boys leave school in the same parlous state mathematically.

Taking all subjects together, between 1951 and 1979, the proportion of girls achieving grade C or above in GCE O-level and CSE
examinations has been higher than the corresponding proportion of
boys. A similar situation has existed for GCE A level in a large
majority of the years. But in individual subjects the picture is
not as consistently cheerful for the girls. Keeping their supremacy
in subjects such as languages and religious knowledge, they lose it
for the sciences, including biology, and for mathematics. For example,
in 1979, 52 per cent of the girls entering for O-level mathematics
achieved grade C or above compared with 60 per cent of the boys.

Table 3   Percentages of sixth-form pupils with A-level grades in
           mathematics leaving in 1979

Grades	Boys		Girls		All	
A	15.4)		10.1)		14.0)	
B	14.6)	42.2	16.1)	37.5	15.0)	41.0
C	12.2)		11.3)		12.0)	
D	15.0		16.0		15.3	
E	16.2		19.5		17.1	
Other	26.6		27.0		26.6	
	100.0		100.0		100.0	

Source: Department of Education and Science, *Statistics of Education
          1979*, Vol. 2 (HMSO, 1981).

It has been suggested that the higher overall achievement of girls at
O level is because of their greater maturity at this age and the higher
overall achievement at A level is a result of the smaller and more
select group of girls who stay on in the sixth form.[12] This latter
point is supported by the fact that the differences in overall achievement have virtually disappeared as the numbers entering for A levels
have become more equal. However, as at sixteen-plus, the pattern of
achievement in mathematics does not follow the all-subjects pattern
(see Table 3). The proportions of boys and girls are comparable
except at grade A and grade E.

Since female achievement in mathematics (together with the sciences)
does not follow the overall pattern of pass rates, explanations for the
imbalance in boy and girl successes must lie elsewhere. Concern and
interest have been expressed recently about the possible effect of the
style of examination on performance.[13] It is thought that multiple
choice objective tests (compared to written questions) penalize girls.[14]

If this is proven, it will have serious repercussions for many subjects.

Table 4 shows the numbers of male and female mathematics graduates teaching at primary and secondary levels in 1975 and 1979. These figures include BEd graduates with mathematics as their main subject.

Table 4   Full-time teachers who are mathematics graduates

| | | Year ending 31 March | |
		1975	1979
	Men	131	106
Primary	Women	174	165
	All	305	271
	Men	4804	5455
Secondary	Women	2296	2668
	All	7100	8123

Source: Department of Education and Science, *Statistics of Education 1975* and *1979*, Vol. 4 (HMSO, 1977 and 1981).

Women graduates predominate at the primary level but these are a very small proportion of all primary teachers. The majority of primary teachers are female and most of these have no high level qualification in mathematics.

If girls identify at all with their female teachers, then this identity will often include a lack of competence in mathematics and anxiety about number work. Unless the girls have a strong motivation to persist in the subject, the foundations for a future withdrawal will be laid. The situation is exacerbated at the secondary level by the predominance of male qualified teachers. In a given school, little sex-role identification in mathematics may be available for the girls.

The situation for all pupils is worrying when the data in Table 5 are considered.

Table 5 Estimated percentages of mathematics curriculum suitably*
         staffed in secondary schools

% of curriculum suitably staffed	Estimated number of schools
0	152
1 - 10	27
11 - 20	49
21 - 30	260
31 - 40	250
41 - 50	543
51 - 60	796
61 - 70	780
71 - 80	548
81 - 90	528
91 - 99	199
100	241

Source: Department of Education and Science, *Secondary School
         Staffing Survey* (DES, 1977).

While many mathematically unqualified teachers are able to cope well
with school mathematics, it is difficult for a teacher who is lacking
in confidence or competence to impart encouragement and confidence to
any pupil.

Further investigations are needed to determine the exact nature
and causes of the trends outlined in this chapter. The work of the
Cockcroft Committee may be helpful here.[15] At present one fact is
certain: girls are performing differently from boys in mathematics -
less well overall. What can be done about this? Are the boys
performing adequately anyway? Can educational research provide an
informed and realistic base for practising teachers to work from in
solving these problems?

---

* 'Suitably' = teachers with a good or acceptable mathematics
  qualification.

3. PUPIL ABILITY, ATTAINMENT AND ATTITUDE: A SUMMARY OF RESEARCH

Ability and attainment

An overview

Sex differences in mathematical performance have been a matter for debate for over a hundred years.[16] A number of hypotheses have been developed to try to explain the poorer achievement, on average, of girls. They have been quoted so often it is difficult to know whether they are supported by hard evidence or have gained credence from familiarity. It is important to separate fact from myth.

A comprehensive survey by Maccoby and Jacklin[17] of the research shows that a number of sex differences related to learning can be fairly firmly established.

Girls excel in verbal ability. This finding is consistent with girls performing better in written examinations than objective tests.

Boys excel in visual-spatial ability. This is found consistently in adolescence and adulthood but not in childhood.

Boys excel in mathematical ability. At the pre-school level both sexes are similar in their acquisition of number concepts. Throughout primary school there is no overall difference but girls tend to show a slight superiority in computation whereas boys are ahead on tests of arithmetic reasoning. From the age of twelve-plus, boys' mathematical skills increase faster than girls'. Girls have greater difficulty in the formulation of mathematical or arithmetical problems from a verbal description of the situation.

In contrast, a number of popularly held beliefs are not substantiated by the evidence:

Girls are not better at rote learning and simple repetitive tasks; boys are not better at tasks which require higher-level cognitive processing and the 'inhibition of previously learned responses'. However, boys do seem to be superior at tasks which require the production of solutions to novel situations; this could be a result of early experiences.

Boys do not have a greater analytic ability.

Boys do not have greater achievement motivation. Girls' lower educational attainment is often attributed to lack of achievement motivation - that is, an inherent sense of the necessity of being successful in the tasks attempted. However, many research tests have shown no overall difference. This result is surprising, contradicting personal experience and current views.[18] The differences in actual and expected results may reflect differences in the nature of male and female goals.

A review of research, commissioned by the Cockcroft Committee, discusses in detail other factors related to the teaching and learning of mathematics, such as memory, concept formation and development, classroom behaviour of pupils and teachers, development of general strategies of learning, and curriculum development and evaluation.[19]

It is clear that there are consistent differences between the average scores of girls and boys on tasks designed to measure mathematical ability. However, there is always a considerable overlap in the distributions of scores and the variability of boys' scores is always greater than that of the girls - the highest and lowest scorers are boys. Frequently, over 30 per cent of the girls' scores reach or exceed the median score of the boys. This means that conclusions drawn from information about mean scores provide no information about individuals of either sex. There will be many girls who can equal or surpass their male peers in some or all of the above qualities.

Mathematical ability is not a unitary factor[20] and there are a number of ways of approaching most mathematical problems. There is considerable evidence to suggest that boys and girls approach the solution of a mathematical problem in different ways. If they do, should they be taught by different methods? Apart from the social and general educational undesirability of segregation, studies which have tried to match teaching methods to specific aptitudes and learning styles have so far produced disappointing results.[21]

Early experience

It has been suggested that the intellectual development of girls takes place primarily through linguistic channels while in boys non-verbal skills play a more prominent part.[22] Boys and girls have also been seen to differ in exploratory activity. Brierley found that boys of three to five years are actively exploring and manipulating the environment, while girls become increasingly skilled in verbal and social functions and occupied in sedentary activities such as crayoning, cutting out and plasticine work.[23] Delamont suggests that sex divisions in early learning are accentuated by the expectations of parents and by the toys given to pre-school children. 'The world of toys and games offers girls a far more restricted range of roles than it does boys, and the roles offered girls are essentially passive, home-centred, non-scientific, non-technical, and good.'[24]

A major difference between the sexes in primary school is boys' superiority in spatial and mechanical tasks. Since this appears to give a firmer base for mathematics and science at secondary school than girls' superiority in computation, the interaction of young children and their toys could be a crucial factor for later mathematical progress. The effects on mathematical performance of these differential experiences do not tend to appear until secondary level. The girls' superiority at this age in verbal and basic computational skills carries them through the assessments at primary level.

Primary and secondary surveys

The swing against mathematics at about age thirteen, which is stronger for girls than boys, was mentioned in Chapter 2. When boys and girls do choose to study mathematics, do they perform in the same way?[25]

Robert Wood analysed the type of question successfully answered by boys and girls at GCE O level.[26] For most items the success rate was greater for boys than for girls. Boys were better in items involving scale or ratio, including fractions and probability, or spatial concepts, particularly in three dimensions. Girls were better in items concerning Venn diagrams, vectors, matrix multiplication and rotation. The girls surpassed the boys in questions requiring fairly straightforward skills, in which no problem-solving behaviour was required.

Wood, following Cable,[27] attributes the girls' relative lack of success in scale and ratio problems to a lack of development of the concept of 'comparison factors' - that is, numbers which state how one quantity compares with another of the same kind. Cable claims 'comparison factors' are the basis of quantification. If so, then some pupils' problems (especially girls) stem from the lack of an adequate 'feeling for numbers'.

This lack of a 'feeling for numbers' was shown time and again by boys and girls involved in the Concepts in Secondary Mathematics and Science Project based at Chelsea College.[28] It is clear from this study of eleven- to fifteen-year-olds that a large proportion of children do not understand the processes involved in the problems they are asked to solve.

Part of a five-year investigation carried out in Sheffield[29] involved a large sample of twelve-year-old pupils who were examined at the beginning and end of a school year in arithmetic and problem-solving. On the first tests there was no overall significant difference between the boys and girls. Between-school differences were far greater. On the arithmetic test, girls excelled in questions involving integer operations and simple decimal operations; boys in questions on fractions. Most pupils performed badly on the problem-solving test, though the boys did slightly better on average. The context of the problem appeared to influence the pupils' ability to solve it. Both sexes improved their performance in the second set of tests. Problem-solving scores were still low. School differences persisted but marked sex differences were still not apparent in mean scores.

In the comprehensive primary survey of the DES Assessment of Performance Unit in 1978,[30] involving eleven-year-old pupils, girls' average scores were higher in computation (whole numbers and decimals); boys' in questions about length, area, capacity, application of number and rate and ratio. Some similar findings were reported by the Schools Council Project, Primary School Mathematics: Evaluation Studies.[31] The girls in the sample performed slightly better at straightforward computation, while the boys showed slight superiority in understanding the structure of number.

The Assessment of Performance Unit secondary survey (also in 1978)[32] of fifteen-year-old pupils aimed to identify performance on five learning outcomes: concepts, skills, routine applications, problem-solving and attitudes. In general, the mean scores were slightly higher for boys than for girls in the written tests; the gap widened considerably for the practical tests. The girls performed particularly badly if there was an element of problem-solving in the practical test.

For the last ten years, Ruth Rees and others at Brunel University have been examining mathematical difficulties experienced by a wide range of students, including secondary-school pupils.[33] (Recently the work has been extended to the primary sector.) Three points from the many findings of this extensive study are particularly relevant:

(a) Success in most basic mathematical processes correlated positively with general ability. Success in some basic processes, such as those involving ratio or the multiplication of certain decimals, did not correlate with ability but with some so far undefined mathematical ability.

(b) The pattern of errors and lack of understanding was basically the same at all levels - in particular, student teachers and school pupils showed similar problems.

(c) Small changes in the detail of a question produced fundamental changes in pupils' abilities to answer it. This has obvious implications for teaching.

Attitude

A number of surveys have shown interesting differences in attitudes to learning mathematics and in perceived attributes of boys and girls.[34]

In a small primary school survey,[35] more boys than girls liked mathematics for the same reason that the girls disliked it - they found it challenging and difficult. The girls frequently mentioned their fears of getting sums wrong. Teachers were asked to describe girls they felt were good at mathematics. They were seen as mature, independent, fairly extrovert and good at most subjects.

The Sheffield research[36] mentioned earlier, involving twelve-year-old secondary pupils, also included attitude surveys at the beginning and end of a school year. In the first survey, sex differences were small overall compared with school differences, though girls showed less confidence in their ability to cope with mathematics and thought the subject difficult. Both boys and girls saw mathematics as suitable for both sexes and there was no difference in its perceived usefulness. At the end of the year the overall attitude mean score was significantly lower for girls. Measures of anxiety and lack of confidence were increased, though this was also true of the boys but to a lesser extent. Both expressed difficulty. Attitude scores followed attainment scores fairly closely for boys, but not so well for girls. It seems that success in mathematics does not necessarily generate positive attitudes towards it in girls. In fact, a further survey showed that most able girls underestimated their success in mathematics compared with the teacher's estimate, whereas the boys' estimate tended to coincide with the teacher's.

In the final stage of the Sheffield study a questionnaire survey of ten- to fifteen-year-old pupils found that girls' overall attitude to mathematics tended to be more positive than boys' at ten-plus, but much more negative by fourteen-plus, the cross-over occurring at about twelve. Both sexes saw mathematics as equally suitable for boys and girls, and this did not change with age. It is likely that, at the primary level, girls' attitudes to mathematics are influenced by the predominance of female teachers.[37] At secondary level, the majority of teachers are male, especially in the positions of authority.

Bibby's survey of Scottish school-leavers[38] found that most pupils who had avoided mathematics after the second year of secondary schooling did so simply because they did not like it. Girls seemed more likely to relate 'liking' and 'choosing' and to admit 'difficulty' or subject mixes as reasons for not taking mathematics. More boys seemed convinced of the usefulness of mathematics and satisfied with their decision to choose the subject, and thought it had been valuable in applying for jobs or further education. Even well-qualified girls seemed less convinced about the value of their mathematics.

A survey of attitudes likely to influence school attainment, carried out by Corinne Hutt and others[39] on a sample of thirteen-plus and fifteen-plus pupils in mixed and single-sex state schools, revealed large differences in the attributes which boys and girls ascribe to each sex:

Leadership. Most boys thought boys were better leaders than girls, and most girls disagreed (disagreement was stronger in mixed schools).

Intellectual abilities. Girls estimated girls' abilities favourably; under 25 per cent of the boys agreed.

Individual performance. Boys viewed themselves more positively than girls did. Diffidence in girls was greater in the older group.

Competitiveness. Boys were seen as more competitive by both sexes. Younger girls were more prepared to compete than older ones, perhaps indicating an erosion of confidence with age.

Importance of school achievements. Both boys and girls thought it important for boys, but only girls thought that it was equally important for them.

Subject performance. Boys and girls thought most subjects equally well done by both sexes. The view was not so strongly held for science or mathematics, but there was no evidence that girls saw mathematics as a boys' subject. 10 per cent of the girls rated it as their best subject.

Subject preferences. Liking correlated highly with achievement. In mixed schools, girls most frequently chose mathematics as their worst subject, but not in girls' schools. Here the most frequent worst subject was foreign languages, also chosen by boys in mixed schools. Physical sciences polarized in boys' schools - being the most frequently chosen best and worst subject. Best subjects chosen in girls' schools were history and geography. In mixed schools, girls chose English and boys chose history and geography.

Subjects which ought to be studied:
by boys - girls chose languages, mathematics and technical subjects;
boys chose languages and mathematics;
by girls - girls chose languages, mathematics and domestic science;
boys chose domestic science, languages and mathematics.

Job performance. Girls thought most jobs could be equally well done
by men and women; boys that most jobs were done better by men. Girls
thought domestic chores should be shared, but boys saw them as clearly
sex-typed. (These attitudes were more prevalent in mixed than in
single-sex schools.)

Conclusions

Some factors are clearly established:

1. The swing against mathematics at about thirteen, which is greater
   for girls than for boys

2. Attitude to mathematics becomes less positive with age, the
   deterioration being greater for girls

3. Up to twelve-plus there is no statistically significant sex
   difference in average mathematics scores; differences occur in
   the types of item successfully dealt with by each sex

4. Inter-sex variation is less than within-sex or between-school
   variations

5. Girls are less competent at mathematical problem-solving

6. Girls' confidence in themselves mathematically and their expectations of success are less than boys'

7. The nature of the examination influences girls' performance

8. The structure of mathematical processes is more complex than
   was previously thought.

Some factors are unclear or unknown:

1. The value and effectiveness of intervention work in influencing
   pupils' attitudes and achievements

2. The relationship between the nature of mathematical process and
   the sex of the pupil

3. The relationship between differences in play experiences and the
   ability to solve mathematical problems

4. The triggering mechanism for the deterioration in attitude
   towards mathematics.

4. WHAT CAN BE DONE ABOUT SEX-DIFFERENTIATION PROBLEMS?

In view of the many forces working to divorce girls from mathematics, it may seem more economical of effort to accept the situation and concentrate on improving the performance of boys. However, this approach would not only be morally wrong, it would be detrimental to all in the face of our modern technological society. Why should we hinder our national progress by not fulfilling our national potential?

The dual problems of encouraging all pupils, but especially girls, to look favourably upon mathematics and of helping them achieve their full potential may be attacked simultaneously on a number of fronts.

Classroom approaches and teaching methods

The task within the classroom is not simply one of removing all sex stereotyping and bias from textbooks and teaching materials, important though this is. The pupils, especially the girls, must be actively encouraged in mathematics. The role of mathematics in society and its potential role in specific careers must be clearly explained.

Sex differentiation can arise in the classroom in various ways. The following list of questions may be helpful in considering general class-room approaches and mathematics teaching:

* 1. Do we consciously look for sex bias in our classroom practice?

  2. Do we consider that some activities or tasks are more appropriate for boys than girls? Do we allocate tasks on the basis of sex?

  3. Do we have particular expectations for each sex?

* 4. Do we consciously arrange groups of mixed sexes?

* 5. Do we actively encourage resistance to stereotypes?

* 6. Do we consciously try to avoid using sexist language?

  7. Do we tend to talk to the boys when explaining a new idea in mathematics? Perhaps we expect a more ready understanding from them.

---

* Good practice.

8. Do we tend to face the boys when discussing an example or developing a theme? Perhaps we expect misbehaviour to arise in this quarter.

9. Do we tend to ask the boys to provide the answer to a verbal question?

10. Are we concerned that the girls will be upset if their answer is wrong?

11. Do we distinguish on a sex basis between the types of task for which we give praise: commending the girls for neatness of presentation, chiding the boys for the converse; praising a boy for solving a mathematical problem and refraining from comment when a girl does the same?

12. If a boy gets a problem wrong, do we encourage him to try again, offering hints and suggestions, whereas when a girls is in error do we simply provide the answer or pass quickly on to the next question?

Mathematics is different from other school subjects in its initial demands and development. Once basic reading, spelling, vocabulary and writing skills are gained, increments in learning are more gradual. In mathematics, new concepts and frameworks are continually being introduced and the relevance of past learning is not always obvious. For many children there is an initial confusion in mathematics not present in other subjects. Sometimes this confusion is never dissipated. Consequently a positive, encouraging, confident attitude in the teacher is essential. Girls also need frequent and positive feedback in their work.

Pre-school/nursery school

All children should be encouraged to participate in all classroom activities and in activities in the home. There should be no stereotyping imposed on either sex. This is something for both the teacher and the parent to consider.

Girls should be encouraged to use construction kits and three-dimensional toys to develop spatial ability; boys should be encouraged to play games which develop their verbal skills and their hand/eye coordination which can often be poor at this age. The spirit of enquiry should be encouraged in girls.

Primary school

It is essential at this stage that good teaching and classroom practice is always maintained. This practice is thoroughly discussed in the Cockcroft Report.[40] All pupils should be encouraged to participate fully in mathematics lessons. Where necessary to build up confidence, special encouragement should be given to girls to take part and develop investigative skills. They should not be left to get on quietly at the back of the classroom.

Frequent success must also be ensured for all pupils, but again especially for girls. Since girls tend to fear failure, it is essential to use a mathematics scheme which allows children to progress at their own speed and does not place them in undue competition with their peers. Demands need to be made in small 'packages' so that success can be achieved relatively quickly. The child must not experience repeated failure. Judicious use of electronic calculators can be invaluable here. For example, if the multiplication of integers is required to solve a problem, the calculator may be used to obtain the answer once the child has identified the process required. Failure in the whole problem should not stem from failure in carrying out a standard process.

All pupils, except the most able, will benefit greatly if allowed more thinking time when a new topic is introduced. Many pupils find that they have not internalized a given process before being rushed on to the next. Confusion is added to confusion and no sound foundation is ever built.

Secondary school

The ideas and techniques appropriate for pre-school and primary school are equally applicable to the secondary school. Ideas and experiences gained in the early years can be developed and abstracted into formal mathematical thinking. The range of mathematical skills may be extended and more widely applied.[41] The transition from junior school mathematics need not be dramatic; play still has an important role in developing ideas, particularly in the early years at secondary school. Girls will benefit enormously from experience in three-dimensional work, beginning with as much 'play' as possible with solid objects which can actually be cut. This will not only help them to develop manual skills but also to develop abstract visualization skills. This 'play' will help many boys as well. No difference should be made in the play experiences given to girls and boys. Walkerdine and Walden are currently engaged in a three year study of girls and mathematics which will focus on the end of primary and beginning of secondary age groups and the transition between these, and should yield much valuable information about this critical period when girls' expectations of and confidence in their mathematical ability diminishes.[42]

As at the junior level, pupils must be given time to assimilate new concepts. Many boys may be motivated to struggle on to overcome any confusion, but this motivation may be absent for most girls who then give up the struggle. Perceived career requirements are important - thus boys may appreciate that an engineer or a scientist requires a mathematics qualification, but girls whose aspirations sadly lie in differing directions may not share the same career motivation. Ironically, however, even those career paths trodden traditionally by girls, e.g. teaching, nursing, social work, require a mathematics qualification.

If a new topic is taught through its applications and the examples represent as many interests as possible, the internalization process may be accelerated. A point of identification may help understanding. This cannot be done with every topic, but if the technique is used whenever possible, pupils will accept the occasional abstract topic more readily. Placing mathematics in the 'human context' will not only make it more attractive to most girls, but help to avoid the problems of 'transfer of knowledge' experienced by both sexes. Basic techniques in arithmetic and algebra can easily be applied to simple situations in business, medicine, the social sciences and public administration. The last two areas provide a fund of examples for statistics.

It is necessary to try to teach the girls to regard failure in the solution of problems as information, not as an indicator of poor ability. Girls should be encouraged to try alternative strategies in the face of difficulty, to concentrate on the task in hand not on themselves.

With the onset of puberty two of the most important influences on girls are the opinion of the boys and the current 'pop culture'. It is also at this stage that girls begin to reject mathematics. If they feel that success in mathematics (and the sciences) is synonymous with a lack of femininity in the boys' eyes, it is unlikely they will improve their position mathematically.[43]

The teacher should try to influence the boys' opinion so that it is an encouragement to the girls and not a deterrent. This does not mean formal discussion periods; any small pocket of time, such as registration, when the boys and girls are together, can be used. It has been shown that discussion can be effective here, but also that it needs care.[44] The fact that girls' currently poorer performance in mathematics is overtly acknowledged may convince them of its truth and depress their attitudes even further.

The essential issues of mathematics should be shown not only to be relevant to all pupils but within the grasp of both sexes.

Teaching materials

Pupils' perceptions of the appropriateness of mathematics may be reinforced by the type of materials used. Many textbooks are highly stereotyped in the types of task assigned to males and females:

> 'Mary goes shopping with her mother...'
> 'John helps his father build...'
> 'A boy rides a bicycle at 10 mph...'
> 'A girl makes a cake using...'

If the teacher's own examples and exercises follow this pattern and restrict mathematical applications to science and engineering, then most girls, and the non-technically minded boys, will assume that mathematics has nothing to offer them.

A number of modern textbooks try to avoid sex-stereotyping by removing all mention of human beings:

> 'A particle moves at 5 kph...'
> 'A coin is tossed...'

This can also be prejudicial to pupils' attitudes as it implies that mathematics has no place in activities involving social interaction, which is of particular importance to many girls.

The following questions may be helpful in considering the suitability of textbooks or other class material:

1. Do girls and boys/men and women consistently have equal representation in:

   (a) pictures
   (b) written examples
   (c) use of personal pronouns?

2. Do they participate equally in physical and intellectual activities?

3. In problems, are girls and boys/men and women shown in a variety of roles, including non-traditional ones?

4. Are girls/women portrayed as competent and decisive?

5. Is the language used non-sexist (for example, police officer instead of policeman, headteacher rather than headmaster)?

6. Are examples and illustrations drawn equally from girls' and boys'/ men's and women's lives and activities?

However, it is important that one does not lose sight of reality when removing stereotyping from teaching materials; a book in which all the bricklayers are female and all the nurses are male would be so much at variance with the children's experiences as to be totally unreal, thus again divorcing the associated mathematics from any usefulness. A balance must be achieved between the avoidance of stereotyping, thus broadening the pupils' horizons, and the creation of totally artificial situations.

There are many mathematics teachers who are aware of the problems of sex differentiation and the problems experienced by girls in mathematics. These teachers spend much time in the production of stimulating and un-biased teaching materials for their own use. It is now vital to ensure that they are supported by the commercially produced teaching aids and are given the opportunity to share their work with colleagues for mutual benefit. The sharing of materials and experiences may be aided by teachers' centres and by the Girls and Mathematics Association (GAMMA). This is a national association which aims to reduce sex differentiation in mathematics teaching and to encourage the participation of girls in the subject.[45]

School organization

The key issue is to avoid using sex as a category in organization. Administratively convenient, it encourages pupils to see themselves in sex roles rather than as individuals.

School organization impinges on all subjects, but may affect some more than others. Streaming or setting is frequently used in mathe-matics, English and foreign languages. A predominance of girls in an openly acknowledged low ability grouping in mathematics will suffice to reinforce the view of both sexes that 'girls cannot do mathematics' and, therefore, should not try.

Teachers should examine the reasons for this type of situation. Is the timetable arranged so that low level mathematics courses are forced on girls if they choose certain options? Do boys have more opportunity to study mathematics?[46] Are girls given enough encourage-ment and advice on their difficulties? Does the school give adequate careers information and advice early enough? Pupils in the lowest mathematics sets may be those with low educational motivation generally, and girls tend to predominate in this grouping. However, it is increasingly important that girls are made aware from the earliest age of the value of a sound basic education.

Mathematics is not learned in mathematics lessons only. Every opportunity should be taken to liaise with those responsible for the mathematics of physics, biology, geography, technical drawing, etc., areas which are immediate examples of the use and practice of mathematics. These are problems to be tackled as part of school policy.

Options and careers guidance

All options must be open to both sexes and boys and girls helped to make informed choices. Careers guidance must be available when choices are being made and both sexes must be encouraged to consider a wide spectrum of possible futures. Separate careers counselling for boys and girls by male and female teachers respectively can foster stereotypes and choices made on a basis of convention only.

The scope of careers education must be developed. Children's horizons can be widened by examples of sex-role appropriate behaviour: female speakers could be invited to describe careers in engineering and management, while male speakers might explain the opportunities available in nursing or primary teaching. Local employers via the schools and industry link schemes, for example, may stimulate interest among pupils. The establishment of purpose and relevance is particularly important in mathematics which is conceptually difficult in parts and can seem so abstract.

With craft options, girls should be encouraged to try woodwork and metalwork if they are interested. A single girl in a craft class should not be made to feel isolated or a freak. A great deal of encouragement is required; humour should be used with care!

The issue is not that every girl must turn her back on traditional female occupations and become a carpenter or a bus driver or an airline pilot. It is that all pupils should be exposed from a very early age to the possibilities which are available. In mathematics, all pupils should be aware of the requirements of various careers and encouraged to achieve these requirements rather than modify their career aspirations because of mathematical difficulties.

Extra-curricular activities

Clubs are particularly useful in involving pupils, encouraging the development of ideas and improving attitudes. It is difficult to find time in class for the many mathematical games available today, and these games form a good basis for a mathematics club. Clubs are also good for developing model building and studying the properties of solids. There need be no element of competition to frighten away the girls. Processes half-understood in class can be explored entertainingly in a relaxed environment.

Parents and society at large

Every opportunity should be taken for discussion, illustrating the importance of mathematics in our technological society, emphasizing career dependence on the subject, and having available specific information on training and careers. In time it will become socially unacceptable to be 'hopeless at maths'.

The Equal Opportunities Commission * has produced some short films and video cassettes showing women in non-traditional roles, which may be of interest to parents.

---

* For further details, contact: Publicity Section, Equal Opportunities Commission, Overseas House, Quay Street, Manchester M3 3HN.

5. HOW CAN TEACHERS CONTRIBUTE TO FURTHER PROGRESS?

All pupil/teacher groups are different, and only broad suggestions can be offered. Teachers should carry out their own experiments on methods and materials, using other people's experience and with the overall objectives in mind. Teachers should encourage positive attitudes in individual pupils and in the general climate of the school.

Much planning in school is influenced by Department of Education and Science policy and the constraints of the public examination system. It may be that a radical reassessment of mathematics provision is needed. The conclusions of the Cockcroft Committee may provide the impetus for such a revision.

Possibly only arithmetic should be compulsory to the age of sixteen, mathematics in depth being reserved as an option. Arithmetic could be used in assessment for entry to further courses or employment. This might well help pupils in building confidence in numeracy. Boys and girls would have to be given equal opportunities and encouragement to follow the more rigorous optional courses.

Teachers can press for alternative courses of development in mathematics education by making their views known through the professional organizations and unions. They can take every opportunity to initiate discussion on improving successful participation of all pupils in mathematics, but especially of the girls.

NOTES AND REFERENCES

1. Department of Education and Science. *Mathematics Counts*, Report of the Committee of Inquiry into the Teaching of Mathematics in Schools (chairman: Dr W.H. Cockcroft) [Cockcroft Report]. HMSO, 1982, Part 1, Chapter 2.

2. Department of Education and Science (see note 1). Part 1, Chapter 4.

3. *Why, What and How?* Mathematical Association, 1976.

4. K.L. Gardiner, J.A. Glenn and A.I.G. Renton (eds.). *Children Using Mathematics*. Oxford University Press, 1973.

5. Department of Education and Science. *Curriculum 11-16 – Working Papers by HM Inspectorate: a Contribution to Current Debate*. DES, 1977.

6. Department of Education and Science. *Survey of Mathematics in Schools 11-18 (1973-74)*. Results of a survey conducted by HM Inspectors. DES, 1975.

7. J. Bibby. 'Sex differentials in mathematics enrolments: sudden death or gradual decline?' Newsletter No 7, Centre for Educational Psychology, University of Edinburgh, June 1980.

8. D. Duckworth and N.J. Entwhistle. 'The swing from science', *Educational Research*, November 1974, 48-53.

9. A. Kelly. *An Unfair Profession: a Review of the Position of British Women in Science*. Centre for Educational Sociology, University of Edinburgh, 1974.

10. Ibid; J. Bibby (see note 7); E.E. Maccoby and C.N. Jacklin (see note 17); E. Saraga. *Girls and Boys: are there Differences in Ability?* Proceedings of a conference 'Girls and Science Education – Cause for Concern?' Centre for Science Education, Chelsea College, 1975.

11. Department of Education and Science (see note 1). Part 1, Chapter 3.

12. R.J.L. Murphy. 'Sex differences in GCE examination results and entry statistics', Paper given at the Organisation in Schools Courses Conference 'Sex Differentiation and Schooling', Cambridge, January 1980 (unpublished).

13. C.A. Dwyer. 'The role of tests and their construction in producing sex-related differences', in M.A. Wittig and A.G. Petersen (eds), *Sex-related differences in Cognitive Functioning: Developmental Issues*. Academic Press, 1979; R.J.L. Murphy. 'Sex differences in examination performance: do these reflect differences in ability or sex-role stereotypes?' in O. Hartnett, G. Boden and M. Fuller (eds), *Sex-role Stereotyping*. Tavistock, 1979.

14. R. Wood. 'Cable's comparison factor: is this where girls' troubles start?' *Mathematics in Schools*, vol. 6, no. 5, 1977, 18-21; R. Wood. 'Sex differences in mathematics attainment at GCE ordinary level', *Educational Studies*, vol. 2, no. 2, June 1976, 141-160.

15. Department of Education and Science (see note 1).

16. M.A. Clements. 'Sex differences in mathematical performance: an historical perspective', *Educational Studies in Mathematics*, vol. 9, 1979, 305-22.

17. E.E. Maccoby and C.N. Jacklin. *Psychology of Sex Differences*. Oxford University Press, 1975.

18. C. Hutt. 'Why do girls under-achieve?' Report from the Department of Psychology, University of Keele, 1979 (unpublished); R.J.L. Murphy (see note 13).

19. A. Bell and A. Bishop. *'A review of research in Mathematical Education'*. Report commissioned by the Cockcroft Committee. Shell Centre for Mathematical Education, University of Nottingham, 1982.

20. W.D. Furneaux and R.M. Rees. 'The structure of mathematical ability', *British Journal of Psychology*, vol. 69, 1978, 507-12.

21. M. Eddowes. 'Learning difficulties in mathematics with particular reference to girls', in M. Eddowes and S.B. Sturgeon. *Mathematics Education and Girls: a Research Report*. Sheffield City Polytechnic and the British Petroleum Company Limited, 1981; E.E. Maccoby and C.N. Jacklin (see note 17).

22. S. Delamont. *Sex Roles and the School*. Methuen, 1980.

23. J. Brierley. 'Sex Differences and Education', *Trends in Education*, vol. 1, February 1975, 17-24.

24. S. Delamont (see note 22).

25. Department of Education and Science (see note 1). Part 2, Chapter 5, paragraphs 208-216 and Appendix 2.

26. R. Wood (see note 14).

27. J. Cable. 'A society for the preservation of fractions'., *Mathematics in Schools*, vol. 5, no. 4, September 1976, 13-14.

28. K. Hart. *Secondary School Children's Understanding of Mathematics: a Report of the Concepts in Secondary Mathematics and Science Project*. Chelsea College, University of London, 1980; K. Hart (ed). *Children's Understanding of Mathematics 11-16*. John Murray, 1980.

29. M. Eddowes and S.B. Sturgeon. *Mathematics Education and Girls: a Research Report*. Sheffield City Polytechnic and the British Petroleum Company Limited, 1981.

30. Department of Education and Science, Assessment of Performance Unit. *Mathematical Development: Primary Survey Report No 1*. HMSO, 1979.

31. M. Ward. *Mathematics and the 10-year-old*. Schools Council Working Paper 61, Evans/Methuen Educational, 1979.

32. Department of Education and Science, Assessment of Performance Unit. *Mathematical Development: Secondary Survey Report No 1*. HMSO, 1980.

33. W.D. Furneaux and R.M. Rees (see note 20); R.M. Rees. 'An investigation of some common mathematical difficulties experienced by students', *Mathematics in Schools*, vol. 3, no. 1, January 1974, 25-7.

34. Department of Education and Science (see note 1). Part 2, Chapter 5, paragraphs 198-207 and Appendix 2, paragraphs B29-B33.

35. G. Weiner. Girls' attitudes towards mathematics: a case study in an Essex primary school. 1979 (unpublished feasibility study for higher degree).

36. M. Eddowes and S.B. Sturgeon (see note 29).

37. Ibid, Chapter 2, page 6.

38. J. Bibby (see note 7).

39. C. Hutt (see note 18).

40. Department of Education and Science (see note 1). Part 2, Chapter 6.

41. Ibid. Part 2, Chapter 9 and Appendix 2, paragraphs B35-B40.

42. This is an extension of their pilot study work on infant and nursery children, a summary of which will be published by London University Institute of Education in the Bedford Way series as 'The Practice of Reason'.

43. Department of Education and Science (see note 1). Appendix 2, paragraphs B21-B22.

44. E. Fennema. 'Women in mathematics: a cross-cultural comparison of status and problems: part 2, USA', paper given at Fourth International Congress on Mathematical Education, Berkeley, USA, 1980 (unpublished).

45. Girls and Mathematics Association, Organizing Secretary, Zelda Isaacson, Department of Teaching Studies, Polytechnic of North London, Prince of Wales Road, London NW5 3LB.

46. Department of Education and Science (see note 1). Appendix 2, paragraphs B23-B24.

FURTHER READING

BIGGS, J.B. *Anxiety, Motivation and Primary School Mathematics*. NFER Publishing, 1962.

DEPARTMENT OF EDUCATION AND SCIENCE. *Mathematics Counts*. Report of the Committee of Inquiry into the Teaching of Mathematics in Schools (chairman: Dr W.H. Cockcroft) [Cockcroft Report]. HMSO, 1982.

FENNEMA, E. 'Women and Girls in Mathematics - Equity in Mathematics Education', *Educational Studies in Mathematics*, vol. 10, 1979, 4.

FENNEMA, E. (ed). *Mathematics Learning: What Research says about Sex Differences* (Mathematics Education Report). ERIC (Education Resources Information Center) Information Analysis Center for Science, Mathematics, and Environmental Education, 1975. Available from ERIC Information Reference Center, Ohio State University, Columbus, Ohio.

FOX, L. 'The Effects of Sex Role Socialization on Mathematics Participation and Achievement', in *Women and Mathematics: Research Perspectives for Change* (National Institute of Education Papers in Education and Work, no. 8). NIE, Washington DC, 1977.

FURNEAUX, W.D. and REES, R.M. 'The Dimensions of Mathematical Difficulties' (Department of Education Occasional Publications, Series 1). Brunel University, 1976.

HART, K. 'Mistakes in Mathematics', *Mathematics in School*, vol. 7, 1978, 4.

HART, K. 'Understanding of Ratio in the Secondary School', *Mathematics in School*, vol. 7, 1978, 1.

HILTON, T.L. and BERGLUND, G.W. 'Sex Differences in Mathematics Achievement - a Longitudinal Study', *Journal of Educational Research*, vol. 67, 1974, 5.

KENT, D. 'Some Processes through which Mathematics is Lost', *Educational Research*, vol. 21, 1978-79, 1.

LOMAX, P. 'The attitudes of girls with varying degrees of school adjustment to different aspects of their school experience', *Educational Review*, vol. 30, no 2, 1978, 117-24.

LUNN, J. BARKER. *Streaming in the Primary School*. NFER Publishing, 1970.

MEIGHAN, R. (ed). 'Sex Roles in Education', *Educational Review*, vol. 27, Special Issue, 1974-5.

RICKS, F.A. 'Teacher Perceptions and Attitudes which Foster or Maintain Sex Role Differences', *Interchange*, vol. 4, 1973, 1.

SLEE, F.W. 'The feminine image factor in girls' attitudes to school subjects', *British Journal of Educational Psychology*, 48, 1968, 212-14.